COUNT TO 50 with SISI

Copyright © Year 2025
All Rights Reserved by **Meryem Pektaskin.**

ISBN
Paperback: 979-8-90190-053-6
Hardcover: 979-8-90190-054-3

Dedication

To my children, who continue to inspire me to tell stories, and to my dear students, especially Olivia, whose curiosity and encouragement reminded me why stories matter.

About the Author

Born in 1993, Alpha Lorenza graduated with a degree in Radio and Television Broadcasting. She holds certificates of honor, a diploma supplement, and several awards recognizing her outstanding achievement in English language studies.

In addition to her passion for storytelling, she is also a songwriter, sharing her original music on her YouTube channel. She currently works as an English teacher, where her students' enthusiasm often inspires her creative work.

Her dedication and love for teaching, writing, and music all came together in the creation of this book—a heartfelt effort to inspire readers just as her students have inspired her.

Hello, I'm Sisi.
I'm twelve years old.
I live with my mother
and my sister.

I go to school.
I learn to
write and read.
I also know
to count the numbers.

I can count to ten.
It's easy for me.
I know you can count to 10.
Count to ten.

One, two, three, four, five, six, seven, eight, nine, ten.

My sister's name is Eva.
She is six years old.

My sister can't count.
She doesn't know the numbers.
She knows the colours.
She loves pink very much.

I learn Math at school.
My teacher teaches us the numbers.
We count to twenty at school.

Eleven, twelve, thirteen, fourteen, fifteen.
Tell me 15.

My mother's name is Alice.
She is thirty-seven years old.
She knows to count.
She knows every rules about Math.

Do you know to count to 37?

Continue to count.
We count the fifteen.
It's 15.

Fifteen, sixteen, seventeen,
eighteen, nineteen, twenty.
Tell teen, teen, teen.

My mother teaches you to count to thirty.
It's 30.
It's easy.

Twenty-one-two-three-four-five-six-seven-eight-nine.
Twenty + numbers.

You can count to thirty now.
You are ready to continue.
Let's continue!

Say thirty.
Thirty-one-two-three-four-
five-six-seven-eight-nine.
Thirty + numbers again.

Say 41.

It's fourty-one.

Fourty-one-two-three-four-five-six-seven-eight-nine.

Remember Fourty + numbers.

It's fifty.

You know to count to fifty now.
It's OK now.

ARE YOU READY?

Write the numbers again.

COUNT TO 50 with SISI

one   1

two   2

three   3

four   4

five   5

Write and color the numbers again.

COUNT TO 50 with SISI

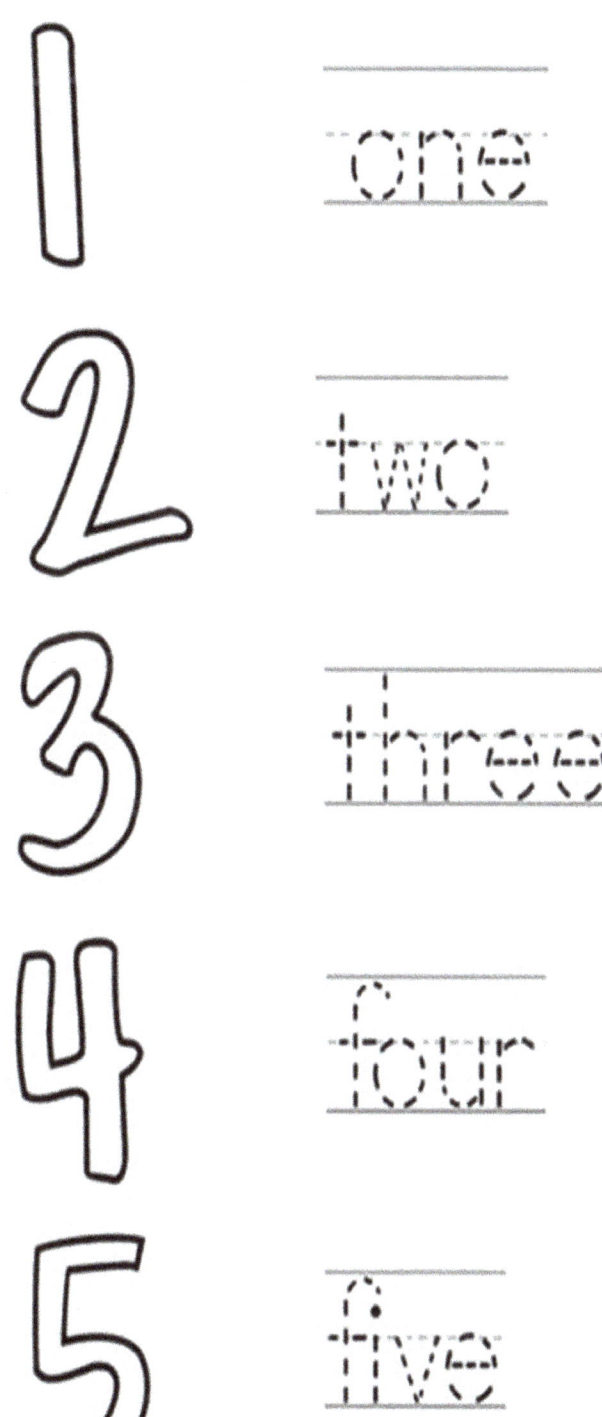

1 — one
2 — two
3 — three
4 — four
5 — five

Match the correct one.

COUNT TO 50 with SISI

One	2
Three	4
Two	5
Five	1
Four	3

Fill the blanks and write the numbers.

COUNT TO 50 with SISI

one	two	three
four	five	six

1   4   3

6   2   5

Count and correct it.

COUNT TO 50 with SISI

 • 4 • Five

 • 5 • One

 • 2 • Two

 • 1 • Four

Color the numbers.

COUNT TO 50 with SISI

4 ♡ ♡ ♡ ♡ ♡

3 ♡ ♡ ♡ ♡ ♡

5 ♡ ♡ ♡ ♡ ♡

2 ♡ ♡ ♡ ♡ ♡

1 ♡ ♡ ♡ ♡ ♡

Write the numbers again.

COUNT TO 50 with SISI

six 6

seven 7

eight 8

nine 9

ten 10

Match the correct number.

COUNT TO 50 with SISI

1	2
3	4
5	8
7	3
8	6
9	10
10	1
2	5
4	9
6	7

GOOD JOB!

Write the numbers again.

COUNT TO 50 with SISI

eleven   11

twelve   12

thirteen   13

fourteen   14

fifteen   15

Write the numbers again.

COUNT TO 50 with SISI

sixteen 16

seventeen 17

eighteen 18

nineteen 19

twenty 20

Match the correct number.

COUNT TO 50 with SISI

Eleven	20
Twelve	14
Thirteen	13
Fourteen	18
Fifteen	15
Sixteen	17
Seventeen	19
Eighteen	16
Nineteen	11
Twenty	12

Read and find the correct one.

COUNT TO 50 with SISI

11
- Eleven
- Elefen
- Eleveen

12
- Tlevee
- Twelve
- Tweleve

13
- Thirteen
- Thirten
- Thirdteen

14
- Fourtteen
- Forten
- Fourteen

15
- Fifteen
- Fiveteen
- Fevten

16
- Sixthteen
- Sixtten
- Sixteen

17
- Seventen
- Seventeen
- Sventeen

18
- Eightteen
- Eghten
- Eighteen

19
- Nineten
- Nineteen
- Ninten

Write the numbers (Before-After.)

COUNT TO 50 with SISI

Before	Number	After
3	4	5
☐	3	☐
☐	9	☐
☐	2	☐
☐	5	☐
☐	7	☐

Before	Number	After
☐	6	☐
☐	11	☐
☐	10	☐
☐	8	☐
☐	1	☐
☐	4	☐

Write the numbers (Before-After.)

COUNT TO 50 with SISI

Before		After		Before		After
6	○	8		16	○	18
4	○	6		14	○	16
7	○	9		17	○	19
5	○	7		15	○	17
2	○	4		12	○	14
6	○	8		16	○	18
1	○	3		11	○	13
3	○	5		13	○	15
9	○	11		19	○	21

Write and color the missing numbers.

COUNT TO 50 with SISI

Write the numbers in the box.

COUNT TO 50 with SISI

21					
22					
23					
24					
25					
26					
27					
28					
29					
30					

Write the numbers in the box.

COUNT TO 50 with SISI

41					
42					
43					
44					
45					
46					
47					
48					
49					
50					

LET'S MAKE AN EXAM!

Count the numbers.

COUNT TO 50 with SISI

5 + 2 = ___	4 + 1 = ___	8 + 1 = ___
2 + 6 = ___	3 + 2 = ___	8 + 2 = ___
3 + 6 = ___	9 + 1 = ___	4 + 2 = ___
3 + 5 = ___	6 + 4 = ___	7 + 3 = ___
1 + 5 = ___	7 + 1 = ___	5 + 5 = ___
3 + 3 = ___	3 + 4 = ___	1 + 2 = ___
5 + 4 = ___	2 + 2 = ___	6 + 1 = ___
7 + 2 = ___	1 + 3 = ___	1 + 1 = ___
4 + 4 = ___	2 + 5 = ___	1 + 4 = ___
1 + 8 = ___	6 + 2 = ___	2 + 3 = ___
2 + 8 = ___	6 + 3 = ___	1 + 9 = ___
2 + 4 = ___	5 + 3 = ___	4 + 6 = ___
3 + 7 = ___	5 + 1 = ___	1 + 7 = ___
4 + 3 = ___	2 + 1 = ___	4 + 5 = ___
1 + 6 = ___	2 + 7 = ___	3 + 1 = ___
3 + 2 = ___	3 + 6 = ___	1 + 5 = ___

Count the numbers.

COUNT TO 50 with SISI

9+6=	9+6=	9+6=	9+6=
7+7=	7+7=	7+7=	7+7=
9+4=	9+4=	9+4=	9+4=
6+5=	6+5=	6+5=	6+5=
9+6=	9+6=	9+6=	9+6=
7+7=	7+7=	7+7=	7+7=
9+4=	9+4=	9+4=	9+4=
6+5=	6+5=	6+5=	6+5=
9+6=	9+6=	9+6=	9+6=
7+7=	7+7=	7+7=	7+7=
9+4=	9+4=	9+4=	9+4=
6+5=	6+5=	6+5=	6+5=
9+6=	9+6=	9+6=	9+6=
7+7=	7+7=	7+7=	7+7=

4+8−2=	4+8−2=	4+8−2=
9+5−10=	9+5−10=	9+5−10=

www.ingramcontent.com/pod-product-compliance
Lightning Source LLC
Chambersburg PA
CBHW041217130526
44582CB00026BA/89